How To Make Money With 3D Printing:

Passive Profits, Hacking the 3D Printing Ecosystem and Becoming a World-Class 3D Designer

Jeffrey Ito

Preface

The Next Industrial Revolution Is Coming

Imagine a future in which people create products, prototypes and art from the comfort of a computer. A future in which distribution of tangible goods means data delivered over the Internet. And a future in which the reduction of harmful emissions, elimination of inventory and introduction of mass customization ushers in a third industrial revolution.

3D printing is the future and the industry is growing rapidly – right now. The single biggest question I encounter regarding 3D printing is how one can capitalize on the growing technology. Having witnessed this problem in dozens of books, magazines, forums and even my own personal conversations, I was compelled to write this book. Most guides I have read are either too technical or theoretical for the everyday person to easily understand, so my goal is to simplify the disruptive technology in a way in which everyone may benefit.

Disclaimer

No part of this book may be reproduced or transmitted in any form or by any means, electronic or mechanical, including photocopying, recording or by any information storage and retrieval system, without written permission from the author. The information provided within this book is for general informational purposes only. While we try to keep the information up-to-date and correct, there are no representations or warranties, express or implied, about the completeness, accuracy, reliability, suitability or availability with respect to the information, products, services, or related graphics contained in this book for any purpose. Any use of this information is at your own risk. The methods describe within this book are the author's personal thoughts. They are not intended to be a definitive set of instructions for this project. You may discover there are other methods and materials to accomplish the same end result.

The information contained within this eBook is strictly for educational purposes. If you wish to apply ideas contained in this book, you are taking full responsibility for your actions.

Copyright © 2015 by Jeffrey Ito.
All rights reserved worldwide.

No part of this publication may be replicated, redistributed, or given away in any form without the prior written consent of the author/publisher Jeffrey Ito.
Jeffrey Ito
United States
http://www.moccidesigns.com

Contents

Contents 6
   Bonus Features 8
What is 3D Printing? 11
   Advantages of 3D Printing 11
   Modern Design Twist 12
   Printing With The Cloud 13
   Eradicate Inventory & Specialize in Made-To-Order 14
   Green Emissions 16
   Customization 17
   Rapid Growth 18
Disadvantages of 3D Printing 20
   Gartner Hype Cycle for 3D Printing 23
   Inspiration 23
   Democracy of Manufacturing 24
   The Open Canvas 29
I.   3D Designs 31
   Create 3D CAD Model 31
   Outsource CAD Design 37
   3D scanners 43
   Digital Marketplaces 50
II.   Physical products 56
   The Long Tail of The Internet 61
   Smart Niching 63
   The Long Tail of Things 65
   Disruption With 3D Printing 66
   Reach Your Niche Market On The Internet 82
   The Power of Free for Big Businesses 86
III.   Build 3D Printers 87

- Democracy of Distribution, Manufacturing and Funding 88
- The RepRap Project 90
- With Great Challenge, Comes Great Satisfaction . 93
- Kickstarting Your 3D Printer 94
- IV.   Share Your 3D Printer 100
- The Shared Economy 101
- 3D Printing Hubs 101
- Expanding The Distributed Factory 113
- V.   Customization as a Service 114
- The Potential Of On-Demand Mass Customization 114
- Case Studies 115
- Partnering with 3D Printing Service Providers . 118
- VI.   Invest in 3D Companies 126
- Putting Money Where My Mouth Is 127
- 3D Printing Stocks That Are Disrupting The Industry 128
- 3D Systems (DDD) 128
- VII.   Teach 132
- Conclusion & Thoughts 135
- About The Author 136

Bonus Features

- 2015 3D Printer Buyer's Guide
- Movies & TV episodes about 3D printing
- The 8 Best 3D Modeling Software for 3D Printing Entrepreneurs
- 3D Scanner & Apps Guide
- The Top 10 Places To Outsource 3D Design Work
- Specialty CAD software
- 3D Printing Services Available Today
- The Top Places Online To Sell Your Physical Creations
- 3D Printers Successfully Funded Through Kickstarter
- The Top 3D Printing Networks
- 3D Printing Stocks

Introduction

It's Sunday morning, the birds are chirping and your first interaction with life is a warm sunbeam gently invading your sleep through an open slit in the blinds. You open your eyes, assess bodily signs, roll out of bed and start your day – worry free. A smile stretches across your face this morning because your bank account is richer than it was when you went to bed yesterday. Your 3D printer is grinding away 24 hours a day creating things you thought up in your dreams.

Sound like a fantasy? Well it's not as farfetched as one would think.

There are many books that will teach you what 3D printing is or how to create your own 3D designs, but people are quick to ask one simple question - **How Can I Make Money With 3D Printing**? If there is one thing I am sure about, it is that 3D printing is a game changer and will soon become a global phenomenon. In the last few years the disruptive technology has become an essential tool for artists, engineers, architects and designers alike. The efficiency & effectiveness with which we are able to rapidly prototype things today using 3D printing means that start-up costs are plummeting. The popularity of 3D printing rises as makers recognize that the technology is dripping with potential. Like any other pioneering technology, early adopters will seek to find ways to capitalize on this emerging industry.

This is not a get rich quick guide. It takes hard work and persistence to start a business. The challenge today is transforming data into physical form with 3D printers. I am here to take the pain out of learning 3D printing and directly teach you how to monetize it in all facets. I will also include many fundamentals of creativity along with basic entrepreneurship and business principles that I feel are necessary for success with 3D printing.

What is 3D Printing?

3D printing is a powerful creation tool. Stereolithography (also known as additive manufacturing or simply, 3D printing) is the original process of making three-dimensional objects from a digital file (.stl). The file is created with a 3D modeling program or scanner. As it is sent to the printer, the software slices the design into thousands of horizontal layers. These layers of material are then printed successively until your physical object is created. Magic!

Advantages of 3D Printing

"The victorious strategist only seeks battle after the victory has been won, whereas he who is destined to defeat first fights and afterwards looks for victory." – Sun Tzu

Modern Design Twist

The introduction of consumer 3D printers instantly gave 3D designers the power of manufacturing. Anyone, anywhere can design and then create anything! Consider this – engineers, architects, animators and artists once used **Computer Aided Design (CAD)** programs just for the graphical realization of their ideas on a screen. Computer generated models have been indispensable for creative endeavors such as the architectural, video game and film industries for decades. Now with 3D printing we can print these creations to life! The artisan spirit died with the Industrial Revolution, but now the ability to creatively manufacture our own things means that people can once again create businesses for themselves.

If you are neophyte to 3D printing, most likely you are new to 3D modeling as well. Luckily, the rise of 3D printing has attributed to an increase in CAD innovation. So, programs are being developed with 3D printing in mind such as Google Sketchup and TinkerCAD are easier than ever to use. Not to mention the old heavyweights - AutoCAD, Rhino and Blender - are all as pristine design tools as they have ever been.

Printing With The Cloud

Imagine you wanted to send a friend a hand painted porcelain Oriental vase. Instead of facing the rigors of expensive packaging with extra care for shipping, attaching postage, waiting in line at the post office and waiting days for the vase to arrive, you will be able to seamlessly send the vase to your friend via 3D printer and the cloud. This is the ideal future of distributed manufacturing at work. Every house or business has the potential to become a 3D printing hub for manufacturing. For skeptics this may sound like a scene from The Jetsons, but in due time with a few more steps of technological progress – **the future is here and now**.

The cloud partnered with 3D printing will shred distribution costs for business. Sending a file rather than a physical thing will also eliminate packaging for items. Real time data of 3D printers can be monitored remotely with the Internet, which means the possibility of an even more integrated network of manufacturing systems. Be sure to take advantage of the large, creative open-source movement that has emerged to share code, printers and product designs that anyone can use and modify.

Eradicate Inventory & Specialize in Made-To-Order

Traditionally, large manufacturing processes use expensive equipment with a high throughput that must be maintained to achieve economy of scale and minimize the raw materials costs (also known as **Mass Production**). Mass production is inflexible because it is oftentimes difficult to alter a design or process after a line has been implemented. Also, providing variety for consumers with ever growing tastes is challenging.

In the manufacturing world there are two basic forms of mass manufacturing. Milling machines begin with a blank and subtract material until achieving the desired shape. Injection molding machines start with a mold and then material is heated and injected to fill out the space within the mold. These **push** processes specialize in producing as many quantities as possible for a single design, which leads to material waste and inventory abundance. There are extra costs involved every time a product change or modification is implemented and bottlenecks in the system can occur if one process goes down. On a macro scale, these old manufacturing methods are highly ineffective and inefficient.

With 3D printing, there is a cost of zero for changing your design since modifications are made using CAD. By printing things after the customer has purchased it, businesses are able to terminate inventory and focus on tailoring each sale to the customer. This scenario is an example of a **pull** system. 3D printers have the power to control economies of scale. Although we still have many years until this plays out, it is the path 3D printing innovation is headed toward. Instead of worrying about how to cheaply manufacture designs, product makers can now place their focus on what is important: Product Design.

Green Emissions

The 2000s were a time marked by significant environmental consequence. As we as a society strive to reevaluate our environmental footprint in consideration for the growing population and our childrens' children, sustainable technologies have to be considered. Bearing in mind the vast amount of things that are made everyday in factories, 3D printing has the potential to pioneer sustainable innovation in a grand way! Many businesses today are hesitant to make sustainable changes due to a lack of clear competitive benefit despite corporate social responsibility.

The wonderful merit of 3D printers is its environmental friendliness and lack of carbon footprint. Materials science has attributed to zero waste materials for use with 3D printers. Additionally, many of these materials are recyclable and oftentimes can be reused in the 3D printer. Philosophically, 3D printing hubs will cut down on transportation time and subsequently fuel consumption. The relatively small amount of real estate 3D printers take up will increase productivity between processes in manufacturing plants and increase the likelihood of having one in our homes.

Likewise, in a hypothetical future, by purchasing product designs from an "eMarketplace" of 3D products and printing it ourselves, we have eliminated all of the waste from traditional manufacturing. This includes 100% of packaging materials, energy and transportation costs. In this respect users of 3D printing will enjoy a custom-tailored shopping experience while saving the world.

Customization

Customization and automation – two words that are seemingly difficult to fuse. By combining both the Internet and 3D printing start-ups can now provide **Mass Customization** as a service without spending millions in the process. There are a handful of early adopters who are already integrating 3D printing into mass customization for example, Feetz, which specializes in custom 3D printed shoes. Those with dreams of creating mass customization on demand will have to build their own architecture since the technology and associated software is still relatively primitive.

Quite frankly, the spawn of the Internet era was a conduit for global distribution and social connectivity. Now, 3D printing is introducing a new era of creation for artists, designers and makers. The era of mass distribution is now coming full-circle with mass creation... which means anyone, anywhere can design and then create anything for anybody!

Rapid Growth

The 3D printing sector is experiencing rapid growth as many pundits label 3D printing the next industrial revolution. Right now, **Stratasys** and **3D Systems** control the majority of the industrial market, and venture capital-backed **Shapeways** is the leader in on-demand printing, but the field is getting increasingly crowded as the industry shifts to producing real parts and products rather than just prototypes. There are innumerable ways to use

3D printing technology for business opportunities – most of them probably haven't even been thought of yet. What we are exploring here is the possibility of using a budding technology to create income ideas. And what crazy possibilities there are!

Disadvantages of 3D Printing

The cost and speed benefits of 3D Printing versus "Mass Manufacturing" methods isn't yet great enough to change the way we make products, but the moment 3D printing manufacturing technology exceeds that of traditional mass manufacturing methods will be when 3D printing takes over, full-force. Current complaints with 3D printing include statements such as, "the quality is so bad" or "it takes forever to print" or "only nerds can understand these machines". These problems will eventually become its advantages. This is of course the very essence of modern economics: that things get better and cheaper, and therefore more accessible to the majority as businesses compete for pricing. Many big businesses already utilize 3D printing in some way for design and prototyping advantages. The big question is how will 3D printing be used in our homes and offices?

It is important to note that we are closing in on an inflection point for 3D printing, but we aren't quite near the apex either. Just like the printing press in the 1400s, the steam engine in the 1700s and computing in the 1970s, **the moment of right now is a grand opportunity to master the 3D printing technology**. 3D printing is transforming from a new and wildly hyped, but largely unproven, manufacturing process to a technology with the ability to produce tangible, beautiful, complex and sustainable products. A savvy businessperson will recognize the abundance of opportunity in front of us. Since the growth of 3D printing technology has yet to hit its peak, mastery of the subject early on will prepare those who are willing to be ready.

3D Printing Hype Cycle

The Gartner report 'Hype Cycle for 3D Printing' predicts the future rate of innovation and adoption for 3D printing technology. 3D printing for prototyping will continue to see rapid adoption over the next two years across all industries.

In two to five years, there will be a greater adoption of enterprise 3D printing, nurtured by the continued innovation and use of 3D creation software, 3D scanners and 3D printing service bureaus.

In five to ten years, consumer 3D printers are expected to rise in adoption in homes and schools. Medical devices will offer exciting, life-altering new benefits stemming from the global use of 3D printing technology for prosthetics and implants.

In ten years, 3D printing technology is used by businesses for industrial use. Automated factories and supply chains will prioritize 3D printing. As the costs, size and methods of 3D printing improve, objects on a Micro scale will be easily printed.

"One of eight technologies that will 'creatively destroy' how we do business". – Goldman Sachs on 3D printing

Gartner Hype Cycle for 3D Printing

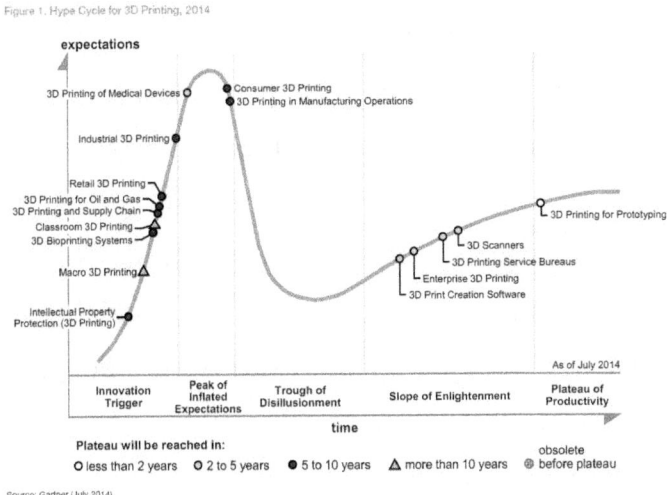

Inspiration

It may be difficult to picture, but corporations have only existed for a few centuries. Imagine you are a spice merchant traveling the Silk Road. Traveling from city to city, you collect new flavors and sell your foreign seasonings to eager buyers. Finding a navigator from Bukhara to Samarkand and through the Himalayas is no longer necessary. This concept of individualized commerce is resurfacing itself. This time with one big difference – 3D printing *combined* with the

Internet. 3D printing will stimulate global economic growth by introducing a new wave of inventive entrepreneurs.

The fascination with 3D printing is more than just another way to manufacture products. 3D printing is a movement that bestows people the power to become makers and create things. Home-based design and creation eliminates unnecessary costly overhead for small businesses and expands imagination into our most coveted tool – the brain. What a win-win.

Democracy of Manufacturing

Notice what happened to the Book, Music, Movie and Video Game industries this last decade? The democratization of media through self-publishing tools has spawned an age where everyone can easily produce and publish as an artist, producer, director or writer. Well, with the Internet & 3D printing the same is soon to change for objects. Before long people will be able to design, print and sell products from the comfort of their own homes.

"3D printing turns digital files into physical objects by building them up layer-by-layer. It gives everyday consumers the power of manufacturing." –Kyle Chayka

Immerse Yourself Into Film

Film is a wonderful learning tool as it is both visual and auditory. Be selective of what images and knowledge you feed into your brain. If you want to dip into 3D printing by first watching and learning then I have included a list of recommend shows and films. On Netflix, **Print The Legend** is a wonderful film that profiles the biggest names in 3D printing and follows their stories and how their businesses have impacted 3D printing technological progress. It is quite an inspiring film. Lisa Harouni's TED Talk, **A Primer On 3D Printing** will prepare you for the possibilities of the swelling manufacturing revolution.

MOVIE/TV SHOW	DIRECTOR
PORTLANDIA Season 4, Episode 9: 3D Printer	Jonathan Krisel
Funny Or Die: 3D Printer	Nick Wiger
Titan American Built TV Series	Titan Gilroy
Heartbeat by You: A Short Film	Lionel Theodore Dean
Print the Legend	Luis Lopez & J. Clay Tweel
Mythbusters: Season 9, Episode 7	Lauren Gray Williams
Lisa Harouni: A Primer On 3D Printing	Unspecified

MOVIE! Featuring: 3D Printing

As if watching actual 3D printers print and listening to people talk about 3D printing wasn't already exciting enough, then wait until you see the end product. The fantastic thing about 3D printing is that high-end prototyping has long been used in the film industry. Creating props and elaborate movie sets has never been easier. Customization of wearable objects to the human form is as simple as scanning the person and designing on the computer.

The Disney Film, **Big Hero 6** features a scene where the protagonist, Hiro Hamada, designs armor for his robot with a Direct Metal Laser Sintering (DMLS) 3D Printer. Even futuristic kids movies are taking advantage of 3D printing. Most Superhero and Science Fiction costumes are constructed with 3D printing because it is faster and more detailed than other methods!

MOVIE	DIRECTOR
The Incredible Hulk	Louis Leterrier
Terminator Salvation	Joseph McGinty "McG" Nichol
Avatar	James Cameron
Ironman 2	Jon Favreau
Harry Potter and the Deathly Hallows Part 2	David Yates
The Muppets	James Bobin
John Carter	Andrew Stanton
The Avengers	Joss Whedon
ParaNorman	Chris Butler, Sam Fell
Skyfall	Sam Mendes
CZ12: A Jackie Chan Film	Jackie Chan
The Boxtrolls	Graham Annable, Anthony Stacchi

The Open Canvas

"Beauty awakens the soul to act." - Dante Alighieri

3D printing is unlike emerging technologies before it in that users of the technology are forced to be creative. The reality of building websites and coding software is that the computer is simply following a series of commands. Enchantingly, 3D printing turns digital files into physical form, layer by layer. Every angle tweak, curvature modification and shape development is another subtle change to how the design will be printed. This seemingly seamless transformation from digital into physical form allows for robust customization with 3D printing. The Internet allows people to do good work and share it with the world. **Thingiverse** is a comprehensive user uploaded resource for finding downloadable 3D things to 3D print. I encourage you to take a peek at the universe of Thingiverse and immerse yourself into the community of things people have designed.

Break All The Rules

People who are crazy enough to think they can change the world are the ones who do." –Steve Jobs

Visit Maker Spaces to get inspired. Ask questions. Find out who your favorite painters, sculptors, philosophers and musicians are and learn from their routines. Go to museums and art galleries. Talk to strangers. Open up your boundaries. Steal other people's work but don't call it your own, make it better. It's not copying – it's remixing and improving. Read a lot of books. Draw the art you want to see. Play the music you want to hear. Develop strong opinions on the world. Do the work you want to see done. Never stop learning.

I. 3D Designs

In and of itself, CAD modeling is an art form. The most creative way to utilize 3D printing technology for profit is to design the "things" that are going to be printed. The process of 3D designing may be digital, but the outcome is physical. In a way this method of creation makes 3D printing the emerging art form of the 21st Century. The truth is artists worldwide are already taking advantage of the depth of design tools and breadth of materials available. 3D printing has even become the norm for prototyping new product designs. As long as 3D printers exist there will be a need for designers, or else what would we print?

Create 3D CAD Model

"Learn the rules like a pro, so you can break them like an artist" – Picasso

The first step towards making money with 3D printing is by creating a CAD model. Now, these drafting programs used to be developed strictly for Engineers and Architects, but with recent improvements towards 3D printing functionality, beginner makers now have plentiful CAD options to choose from, so don't be fearful of the technology.

Look at CAD modeling as an open canvas and you are just getting familiar with your paintbrushes, one brush at a time. Just know that even Picasso used common house paint to make his masterpieces. Mastery of one CAD program makes every subsequent CAD tool learned another design weapon added to your 3D printing arsenal.

The 8 Best 3D Modeling Software for 3D Printing Entrepreneurs

NAME OF SOFTWARE	PRICE	TYPE
TinkerCAD	Free	Beginner
Autodesk 123D	Free	Intermediate
Google SketchUp	Free	Intermediate
Blender 2.72b	Free	Intermediate
AutoCAD Design Suite	$2,760.00	Advanced
Rhinoceros 5	$995.00	Advanced
Zbrush	$795.00	Advanced
Solidworks Professional	$5,490.00	Advanced

I. **TinkerCAD** is the first browser-based 3D design platform. Consequently, it is also one of the great beginner CAD resources. As part of the AutoDesk 123D series, within the web interface there are numerous 'Lessons' to help a newcomer learn 3D design. These lessons range from simple shapes to complex structures. When finished drawing, it is simple to save and send your designs anywhere through the cloud.

II. **AutoDesk 123D Design** is a powerful, user-friendly 3D creation and editing tool. It is CAD software developed specifically for 3D printing and expands upon TinkerCAD with strong sculpting features. For mechanical design, the program has assembly and constraint support. AutoDesk123D is also available on the iPad. Integrate and print directly to shapeways.

III. **Google SketchUp** is an excellent tool with 3D printer support and is a marvelous launch point for learning CAD design. Sketchup is excellent for architectural and geometric models. There is plenty of help online and on YouTube. Be sure to grab the 'Educational' version of Google Sketchup as the professional version will cost upwards of $590.

IV. **Blender** is an excellent tool for modeling textures and art. It is superb for intricate details and perfecting the shape you need. Blender can create high quality renders with lighting and also animate in 3D. Best of all, it is free and open source which means you'll be able to find answers to your most basic questions seamlessly. There is no better alternative to Blender in the hobby sector of CAD modeling tools.

V. **AutoCAD** is the original 3D CAD modeling software and has been Autodesk's flagship program for over 30 years. Consisting of a powerful collection of intuitive 3D design tools, AutoCAD is used and trusted by architects, project managers, engineers and designers. AutoCAD is the top-selling professional CAD design software worldwide.

VI. **Rhinoceros 3D** is professional curve based 3D modeling software based on a mathematical model called NURBS which is used for generating and representing curves and surfaces. Rhino is popular because it is an affordable choice for advanced software.

VII. **ZBrush** is an advanced 3D sculpting tool. It is most prominently used in the gaming and animation industries for its model sculpting and molding tools. ZBrush is exceptional for creating organic and imprecise objects such as abstract textures or human features.

VIII. **Solidworks** is a great tool for Engineers with its parametric and surface modeling features. The software is significant for mechanical parts and exact dimensional tooling. An engineer would find the scripting and parts analysis capabilities very useful. Solidworks is an intricate tool for creating precise and accurate designs.

The CAD program chosen should depend on your design purpose. A beginner will likely opt for

TinkerCAD, **AutoDesk 123D Design** or **Google SketchUp**. An artist is more likely to be in tune with **Blender**, **Rhino** or **ZBrush**, while a product designer will likely utilize **AutoCAD** or **Solidworks**.

Choose your digital weapon wisely.

Outsource CAD Design

Don't feel like CAD designing? Are you more of a manager type? I highly recommend taking the time to learn how to design 3D models, but if you aren't passionate about designing, well, you can always recruit someone else to do it. The learning curve for most advanced CAD software is very steep and modeling can take hours even for a professional. Outsourcing CAD work makes sense. Of course, the software they use will depend on your design needs, but outsourcing creative work for business can be inexpensive if you know where to look. Although one may be a novice to 3D design, your designer doesn't have to be. There are many options available when it comes to **finding a CAD designer**.

I recommend first researching the best CAD software program within your **niche** (I discuss niches in-depth in Chapter 2) and finding a designer well versed with it. Design oriented fields such as the toy, clothing, jewelry, furniture and art industries will perennially have demand for new styles. Digging deeper it is no coincidence that these industries have their specialized CAD programs for designers. There is a specialized program for just about every design field and within each program is a community of other designers willing to help you out.

Here are the most significant specialty CAD programs and their uses:

Specialty CAD Programs

NAME OF PROGRAM	SPECIALTY	URL
Mudbox	Digital Painting and Sculpting	http://www.autodesk.com/products/mudbox/overview
Marvelous Designer	Clothes	http://www.marvelousdesigner.com/
ArtCAM JewelSmith	Art and Texture Design	http://www.artcam.com/index.asp
CityEngine	Cities & Urban Architecture	http://www.esri.com/cityengine
Sweet Home 3D	Architecture	http://www.sweethome3d.com/
CounterSketch Studio	Jewelry	http://gemvision.com/global/css/
Matrix	Jewelry	http://gemvision.com/matrix/
Jewelry CAD Dream	Jewelry	http://jewelrycaddream.com/
Zbrush	Digital Painting and Sculpting	http://pixologic.com/zbrush/features/overview/
Sculptris	Digital Painting and Sculpting	http://pixologic.com/sculptris/

Rhinoceros 3D	Digital Painting and Sculpting	http://www.rhino3d.com/
Geomagic Claytools	Sculpture, Jewelry, Fine Arts	http://geomagic.com/en/products/claytools/overview
Geomagic Freeform Plus	Figurines, Clothes and Design	http://geomagic.com/en/products/freeform-plus/overview
RhinoGold	Jewelry	http://www.rhinogold.com/
RhinoJewel	Jewelry	http://new.rhinojewel.com/
Moment of Inspiration (MoI)	Shapes, Accessories, Decorations	http://moi3d.com/
Zmodeler	Toys	http://www.zmodeler3.com/

Learn everything you can about your niche, find the best CAD software for your niche and create your 3D printed "it".

I used oDesk to outsource professional jewelry design work to CAD designers from Armenia and Ukraine who have been using

GemVision Matrix and **RhinoGold** for 10+ years – a level of performance that would have taken me years of learning and practice to reach. Pay attention to your niche and which countries & cultures have experts for that niche. Developing a relationship with your designer is just as important as evaluating their skills especially since there is no face-to-face contact.

There are minimal reasons to be weary of outsourcing design. Foremost, mastery of the English language is irrelevant for CAD work. Not only is 3D design taught in schools around the world, but also oftentimes the premier designers are the ones who actively learn and teach themselves. The free nature of most CAD programs today lends to the growing global talent pool of designers. Here is a list of resources for finding 3D modelers, designers and drafting contractors.

The Top 10 Places To Outsource 3D Design Work

NAME OF THE SITE	URL	WEB TRAFFIC (visitors per month)
Fiverr	https://www.fiverr.com/	1,761,299
Elance	https://www.elance.com/	1,209,645
Odesk	https://www.odesk.com/	894,134
Freelancer	https://www.freelancer.com/	487,297
Outsource2india	http://www.outsource2india.com/	231,330
Guru	http://www.guru.com/	159,306
IT Match Online	http://www.itmatchonline.com/	140,760
Cad Crowd	http://www.cadcrowd.com/	72,570
True CADD	http://www.truecadd.com/	57,840

| XSCAD | http://www.xscad.com/ | 48,420 |

3D scanners

So you've just graduated woodworking class. And you've built your first piece of furniture – a loveseat bench. Say you wanted to mass-produce this bench, as a dollhouse toy. Well, there are two options. One is to redesign this bench using a CAD program and hope that the finished design is exactly like the physical wooden version you made. Another option is to scan the three-dimensionality of the loveseat bench, creating a design file in the process, which is then scaled into your chosen size and printed on demand.

The wonderful characteristic of 3D scanners is that they give non-creative savvy types a chance to utilize 3D printing within their business. It eliminates the need for creative design work. With one of any object, duplicating it with a 3D scanner is literally as simple as scan and print. Imagine designing a pirate action figure and duplicating it by the thousands to create a menacing pirate army.

It is easy to see how counterfeiting of items can make physical piracy a problem in the future. Just like the media industries before it, there are challenges that will be faced when it comes to intellectual property protection of 3D files. Not to be a contrarian but there are countless reasons to be against intellectual property and patent laws when it comes to innovation. Let us use 3D printing technology for good, so we can avoid costly litigation and restrictions on sharing creativity. Copyright protection for digital files is unlikely to become a serious issue until 3D printing is the norm for physical manufacturing for big businesses. Until this shift occurs, makers have the freedom to download whatever they want and print whatever they like.

The following 3D Scanning Applications are available on Android and iOS devices. The iSense and itSeez3D apps are compatible with attachable scanners on a tablet.

3D Scanning Applications for Download on iOS and Android

Name 3D Scanning App	URL	PRICE
Trimensional: 3D Scanner for iPhone	http://www.trimensional.com/	$0.99
Trnio	http://www.trnio.com/	FREE
123D Catch	http://www.123dapp.com/catch	FREE

3D Scanner Application And Devices (Combined)

Name 3D Scanning App	URL	PRICE
iSense 3D Scanner	http://cubify.com/en/Products/iSense	$499.00
itSeez3D ios app with Structure Sensor	http://itseez3d.com/	FREE App; $379.00 Structure Sensor

A 3D scanner is a device that analyses a real-world object or environment to collect data on its shape and possibly its appearance. They work by placing the object onto the scanner while the technology reads the physical layout of the object and transfers this data for modification in digital form. From beginner to professional, there is a 3D scanner to fulfill everyone's needs.

High Quality 3D Scanners

Name of 3D Scanner	URL	PRICE
The Peachy Printer	http://www.peachyprinter.com/	$100.00
3D Scanner HD	http://www.nextengine.com/	$2,995.00
Artec Eva™	http://www.artec3d.com/hardware/artec-eva/	$13,700.00
Artec Spider™	http://www.artec3d.com/hardware/artec-spider/	$15,700.00
DAVID-laserscanner 3.10	http://www.david-3d.com/	$2,320.50
Desktop 3D Scanner	http://cad-scan.co.uk/	£899 + VAT
Fuel3D Handheld 3D Scanner System	http://fuel-3d.com/	$1,250.00

Geomagic Capture®	http://geomagic.com/en/products/capture/overview	$14-900.00-$24,900.00
GO!SCAN 50 3D	http://www.creaform3d.com/en	$ 28,000
HDI Advance 3D Scanner	http://www.lmi3d.com/	$17,000
Makerbot Digitizer Desktop 3D Scanner	https://store.makerbot.com/digitizer	$799.00
Rubicon 3D scanner	http://www.rubitech.org/	$550.00
Sense 3D Scanner	http://cubify.com/en/Products/Sense	$419.00

Once you have your design model, I highly recommend using **NetFabb** and **Meshlab** for cleaning up your .stl file models, removing holes, measuring and scaling for 3D printing. Both programs offer free versions and will prevent printing errors before sending your file to a 3D printer. Now your 3D model is designed and ready to print!

Give Away Your 3D Designs And Collect Donations From Your Raving Fans.

"I wanted to develop my lucid dreams tridimensionally." –Joshua Harker

Notoriety is earned from a fan base. If you live life as an artist or designer and crave a loyal following, consider giving away your designs and collect donations instead. Joshua Harker is a 3D artist that has had great success sharing his works. Joshua has created multiple kickstarter projects for his imaginative 3D works of art. Now his website is a launch point for the media to profile him. An artist can either create his or her own website or use a service like **makershop.co** to set up a fan base. If you are talented, this method generates buzz and people will come back and check for new designs week after week. Your supporters will appreciate the free work you have created for them and are more than likely to donate a few dollars to their favorite 3D designer to keep the art coming.

Sell Your Designs

Joe loves painting. Some weeks he can make a decent living off of selling one or two of his minimalist nature-inspired pastel paintings. And if he is lucky, he won't have to wait until after death to see his works of art hung up in a museum. But most of the time he is busy waiting tables to make ends meet. Rest assured as a CAD designer you can ease your starving artist worries. The ability to send and receive files on any device through the cloud means you can unveil your 'Mona Lisa' on DropBox to close friends while on vacation, share it immediately once you step foot off the plane on Social Media for likes, comments and shares (read: social proof) and send it to your 3D printer at home to be ready just when you arrive home. This file won't be an original; it will be downloaded and printed thousands, possibly millions of times. That's perfectly fine because you want the world to see your work. The limitless reach of the Internet and the prodigious connectivity of human communication has replaced the bazaar of yesterday.

The following is a list of 3D printing digital marketplaces where you can view and share your CAD files to eager makers as 3D printed objects. There is no monetization aspect to these digital marketplaces but they are great creative resource for ideas and an awesome place to share your designs and interact with the community.

Digital Marketplaces

MARKETPLACE	URL
Thingiverse	https://www.thingiverse.com/
123D Gallery	http://www.123dapp.com/Gallery/
GrabCAD	https://www.grabcad.com/
3D Warehouse	https://3dwarehouse.sketchup.com/
SketchFab	https://sketchfab.com/
Cube Hero	https://cubehero.com/
MakerShop	http://www.makershop.co/
Repables	http://repables.com/

Sell CAD *as* 3D printed objects

With 3D printing as a creation tool, CAD designers essentially make digitally transferable physical objects. Customers can buy these digital files, but instead of receiving a file to open on a computer, the buyer receives the digital design in physical form in their mailbox at home. The ability to turn bits into real objects is power! Companies like Shapeways, iMaterialise, Kraftwurx and Sculpteo offer a full range of 3D printing services including: uploading your design, 3D printing without a printer in many materials and the ability to open your own shop. The beauty of the system is that they will print, ship and take care of customer service for you. A CAD design entrepreneur only needs to worry about design and marketing.

Shapeways

With Shapeways, designers are able to not only view their designs using its built-in 3D feature amongst a range of materials, but also sell their designs in a shop hosted by Shapeways. For each design, the shop owner decides the printing materials and the markup amount. The best aspect of selling your designs on Shapeways is that they do the technical calibrating, printing, polishing, shipping and customer service for you, so all you have to do is sit back and let the money accumulate in your account.

Sculpteo

On Sculpteo makers can fabricate thousands of objects without investing in an expensive 3D printer. The service includes dedicated support, volume pricing and white label services. The shop works similar to the other 3D printing services where you upload your design and choose your material to sell. In your account information a default profit sharing percentage is chosen and applied to your designs for each sale.

iMaterialise

iMaterialise offers many materials and finishing techniques. Designers can open their stores on the website in order to show and sell designs. Similar to the Shapeways business model, designers decide on the fee they want to add on each 3D print on top of the iMaterialise cost to print. After linking your paypal account and setting the photos and description, your shop is good to go.

Kraftwurx

Kraftwurx is hailed as a platform for 3D printing in the cloud. On the service not only can you upload and print your 3D models, but also you can publish your models as products and sell them as a Kraftwurx shop. In addition, Kraftwurx makes opening an online store easier with its digital factory system. Their dashboard is highly advanced and viewing statistical metrics and managing orders and payments has never been simpler. With their Cocreation system, you don't need to have a 3D model to make something – just an idea.

Sell CAD designs for 3D printing

Those who want to focus on selling CAD designs as files with 3D printing functionality have options. This choice caters to a specific crowd of people who are looking for a repository of 3D models not for movies or animation, but just for 3D printing and creation. It is selling files that are made to be 3D printed, but not as 3D printed items. I have condensed this unique list of options for 3D prints to the following:

MARKETPLACE	URL
My Mini Factory 2.0	http://www.myminifactory.com/
YouMagine	https://www.youmagine.com/
Cults3D	https://cults3d.com
PinShape	https://pinshape.com
Threeding	http://www.threeding.com/
Cuboyo	http://www.cuboyo.com/

Sell CAD designs as 3D files

If you find that you particularly enjoy 3D designing creative convoluted objects and want to see your creations on something bigger then consider starting or joining an animation or video game production studio. If you're an artist, experiment with materials and hold exhibitions.

Here is a list of CAD design marketplaces where you can sell your 3D designs. The files do not have to be optimized for 3D printing. The listing can be a 3D render of anything and everything. One exception is **CGTrader** where it is possible to sell 3D renders along with 3D models for 3D printing.

Sell CAD designs

MARKETPLACE	URL
Creative Crash Beta	http://www.creativecrash.com/
DS 3DVIA	http://www.3dvia.com/
TurboSquid	http://www.turbosquid.com/
Falling Pixel	http://www.fallingpixel.com/
3DOcean	http://3docean.net/
3DExport	https://3dexport.com/
The3dStudio	http://www.the3dstudio.com/
Creative Crash	http://www.creativecrash.com/
3D Burrito	http://3dburrito.com/

II. Physical products

"The creation of something new is not accomplished by the intellect, but by the play instinct arising from inner necessity. The creative mind plays with the object it loves." –Carl Jung

Of course, 3D printing means turning files into physical objects. Animators and digital artists have been creating three-dimensional models for years. Once you procure your 3D design either by designing it yourself, downloading from a digital marketplace, 3D scanning or outsourcing the work, then it is time to print!

A painless way to print is with a home-based 3D printer. I have included a 3D printer buyers guide in Chapter 4 for those eager to start fresh with a new 3D printer. If you don't have a 3D printer, never fret. This is an easy problem to solve. There are many options online to outsource 3D printing. In an effort to present the best possible options for 3D printing your designs – trust me there are many – I have whittled the list down to the five that provide the best service, materials and price.

The Best 3D Printing Services Available Today

Shapeways

Shapeways is the premier online 3D printing service in the world. The New York City based behemoth boasts a diverse selection of 22 materials, the largest of any service on this list, ranging from plastics and ceramics to gold and platinum. Uploading & viewing your design, choosing material and printing is a breeze. Having already 3D printed millions of things shapeways continues to see massive growth. Customer service for technical issues is responsive and I even received personal emails from the staff offering assistance with my model and business. The 3D printing service is incredibly easy to use and recommended for all skill levels.

iMaterialise

iMaterialise is a Belgian 3D printing service that offers a selection of over 17 materials including precious metals and ceramics. Their creation corner comes in handy for those with a design in mind but want to learn more about how to make it. With over 70 combinations of materials and finishes for both individual and personal applications, iMaterialise is a high quality 3D print service with over 20 years of experience. iMaterialise offers a wonderful experience in terms of uploading, adjusting and purchasing your models. If you can overlook the higher shipping cost and extended shipping time, iMaterialise is a high quality 3D printing option.

Sculpteo

Sculpteo is a France based 3D printing service that serves makers in both North America and Europe. The service is fast, affordable and high quality. There is a wide selection of materials offered with in-depth documentation for each. Business owners should look into Sculpteo's Cloud

Engine for the possibility of adding a business factory in the cloud powered by their 3D printing resources. Sculpteo is an excellent 3D printing service provider with a wide variety of materials and a stupendous choice for printing and incorporating design apps.

Cubify

Cubify is the consumer division of 3D systems which sells home 3D printers and also offers a print-on-demand service. To access the 3D printing service, users must navigate to the MyCubify section and register. The overall site looks slick, however the materials selection, quality and pricing indicates that the Cubify 3D printing service is still in its early stages of development. With 3D Systems behind the development, the service is sure to become a 3D printing service staple. Cubify has already started offering their own line of 3D printed toys, clothes and technology, including watches, so it is only a matter of time until 3D Systems has a high quality 3D printing service.

Ponoko

Ponoko is a New Zealand based 3D printing service and one of the first manufacturers to use distributed manufacturing and on-demand manufacturing as a business model. They offer laser-cutting, 3D printing and CNC cutting. Ponoko has access to fabrication machines all over the world and uses distributed manufacturing to supply their customers quickly with prints. In this respect, makers are not limited to just 3D printing but have a wide breadth of machines to help shape and form their projects.

Ponoko is a New Zealand based 3D printing service and one of the first manufacturers to use distributed manufacturing and on-demand manufacturing as a business model. They offer laser-cutting, 3D printing and CNC cutting. Ponoko has access to fabrication machines all over the world and uses distributed manufacturing to supply their customers quickly with prints. In this respect, makers are not limited to just 3D printing but have a wide breadth of machines to help shape and form their projects.

3D Hubs

If none of these 3D printing services sound appealing to you due to their lengthy delivery times (a couple days to a few weeks), then 3D Hubs may be the answer. With 3D Hubs the maker gets to choose amongst their local selection of 3D printers to print their creations. In my case, I used local MakerBots for prototyping small toys. I was also able to create high quality, fully castable wax molds for jewelry from a local SolidScape T76 printer. For an iPhone case side project I've used a sandstone 3D printer from a local Maker Shop. Your mileage with 3D Hubs will vary depending on what city you are in and the distance of the nearest 3D printers. Either way, even if you are not interesting in using someone else printer, 3D Hubs is a wonderful community of makers that encourages a worldwide network of 3D printing.

The Long Tail of The Internet

"3D Printing Will Disrupt Every Field It Touches"
– The Economist

For the greater part of the last century, the selection in stores had always been limited by shelf space. This means inventory was filtered primarily by popularity and location. Hardly a convenience when the last copy of the book you have been looking for is in Brooklyn, New York and you are in Kissimmee, Florida. The products on store shelves are influenced by economic demands and large corporations rather than consumers true needs.

Luckily the Internet has a limitless shelf space. The saturation of websites over the years has organically created and identified true niche demand, while the searches people make on Google and Amazon are direct representations of customer interest. The time of the blockbuster monopoly is gone. Huge corporate advertising campaigns and word of mouth no longer tell us what is "in". Collectively, we can easily discuss and decide trending content for ourselves on social media instead of depending on critics. Distribution of media – movies, music and games – has turned digital. We are entering the age of digital distribution where everyone can create, share and find what they want, and it is only a matter of time until physical goods follow suit.

Smart Niching

"What man actually needs is not a tensionless state but rather the striving and struggling for a worthwhile goal, a freely chosen task." – Viktor Frankl

So now you know how to design 3D models and where you can 3D print them. Many will use 3D printers for creative reasons, but those who want to make money by printing objects should focus on niche items. Why should you pick a niche? I'll explain. Creating demand is difficult, but filling it is easy. Instead of making products first and hoping they sell later, find a market with high demand and low supply and create the product they are looking for. I use the **Google Adwords Keyword Tool** to find search volume and explore similar keywords with comparable search traffic. The keyword tool will tell you the amount of competition present online and the amount of global searches a month for a given phrase. Filling demand is easier done with niche products even if it is searched only a few thousand times a month. If you are the only provider of that niche then chances are those few thousand will find you!

Remember that 3D printers only print one at a time. The markets with the best potential to be disrupted by 3D printing are the small-run, high quality, customized markets. Limited or high quality personalized items tend to have larger profit margins anyways. By utilizing the advantages of 3D printing, makers can also create less work for themselves. Instead of envisioning creating an economy of scale with 3D printing, seek to fill the holes of niches that an economy of scale creates.

"It is a good time to think about how something that is as old as the fashion industry can be redefined with new technologies." –Huang

Here are a few specialized 3D printing services to consider when printing your creation:

JewelDistrict

JewelDistrict is an online marketplace for designers to create and share jewelry designs and have them realized with 3D printing. The service is ideal for jewelry making and stone setting. From Amethyst to Emerald to Sapphire, there is an essential collection of gemstones for your jewelry design.

Matter.io

Matter.io lets you start with a series of photos (or sketches) and their team of specialists will help you finish your 3D print. Matter.io prints in all of the basic metals and even offers plating options. With a relatively short 5-day turnaround time, Matter.io is a great solution if you are creating metal works with 3D printing.

The Long Tail of Things

You won't be able to offer everything to everyone online. Creating Amazon 2.0 is just not plausible in the middle of the 2010s, so you'll need to take advantage of a need in the market. Just know that there are many niches out there with very little competition. Stray away from the typical products you see on Wal-Mart or Target shelves and think about what they *don't* carry. My recommendation is to research keywords related to your niche. Then narrow it down by adding long tail keywords to your search term and thus your product. People prefer specificity when it comes to solving their needs. For example, instead of producing general one size fits all hats, narrow it down to stylish platinum bucket hats for infants.

"Why not use this amazing technology to give everybody access to manufacturing? 3D printing as way, not to make just models but, to make actual products." –Weijmarshausen

Disruption With 3D Printing

I oftentimes find that people are stumped as to what they can 3D print that other people will find useful. There are many things, in fact, an infinite amount of things you can make. In my eyes, these are viable industries for building your niche with 3D printing technology today. The barrier to entry on almost all of these industries is essentially access to a 3D printer and the imagination.

Household objects

Common household objects are usually ripe for innovation because they are encountered on a daily or weekly basis. Think about the daily obstacles you encounter at home no matter how minute they may be that, if fixed, could make daily life easier. Do you often misplace things? Would a physical invention make something easier or faster? Examples include a toothpaste squeezer, bottle openers, bookmarks, kitchen instruments, business card holders, measuring devices and utensils.

Art

The infinite nature of 3D design and the bountiful selection of materials make 3D printing a wonderful tool for artists, especially for experimenters of new methods. An artist should at least be familiar with 3D design to utilize the future advantages of creation and sharing. Consider creating physical masterpieces with clay or sand and then 3D scanning your work of art, so it can be saved and shared. With an extensive selection of materials available for 3D printing, consider utilizing this key technology of the future to produce your next work of art.

Furniture

Finding comfortable material for 3D printed furniture may be a challenge, but the possibilities for exquisite furniture design exist in all scales. The physical design nature of furniture lends itself well to 3D modeling and 3D printing. With 3D printing, prototyping a new piece of furniture no longer means taking a lesson in wood shop. The list of ideas for furniture includes: beds, tables, chairs, drawers, lamps, armoires, stools, racks, ottomans, cupboards and storage containers. Other possible suggestions are furniture accessories such as lampshades, flowerpots, curtains and hammocks.

Homes

The ability to dramatically eliminate time and costs involved in building a home with 3D printing will transform worldwide construction. The typical American home takes approximately 6 months to build; so improving this time would substantially decrease construction costs and accidental human fatalities. A team of interdisciplinary USC professors and partner research corporations has worked together to develop Contour Crafting, which is larger than life home fabrication that combines robotics, welding tools and 3D printing.

Contour Crafting prints a concrete house within days. For those living in shantytowns, slums or overcrowded situations, 3D printing will provide a cheaper and more sanitary situation. The technology would also be useful for households facing debt and for emergency situations caused by natural disasters. Contour Crafting and advancements in 3D printing and materials science alone will help to dramatically solve the global problem of homelessness.

Prosthetics & Reconstructive Tissues

Prosthetics is a fascinating field that is ripe for innovation. The individualized nature of building a synthetic limb has made providing assistance to everyone difficult. The beauty of using 3D printing to customize prosthetics for those without limbs is attributing in a major way for the medical industry. Printing body parts and healing tissues such as ears, kidneys and noses is the next wave of reconstructive surgery. Implants will look more real and will be easier to implement.

Hospitals will be able to scan people's faces; use facial symmetrical evaluation software and 3D print aesthetically pleasing faces. 3D printing could be a boon not just for the plastic surgery industry but also for those with damaged facial features. Imagine being able to heal a fallen comrade on the battlefield on the spot. An earless person no longer has to live with one ear. These possibilities are just scratching the surface of what 3D printing can do for the medical industry.

Wearable Technology

Any item of clothing we print and put onto our bodies will be customized to the dimension of our bodies. Software combined with 3D printing can create custom fitting shoes and headphones. Our hats will fit the shape of our heads perfectly. Rings and watches will be customized to our limb size. Eyewear will finally match the diameter of our heads. The dental industry has successfully utilized 3D printing for quick molds and braces. 3D printing is the perfect complement for the rise in wearable technology.

Sex Toys

You're going to have to use your imagination to conjure up variations of 3D printed sexual toys. Items for all shapes, sizes and uses will be available. Depending on material the flexibility, friction and hardness will differ. Of course, health measures will need to be taken to ensure these items are safe to insert into orifices. Celebrities, namely porn stars, will be able to scan and duplicate body parts and distribute it worldwide to dedicated perverts and groupies. I can't foresee a future where personalized sex toys aren't printed on demand.

Mobile Accessories

In today's ever-increasing world of interconnectivity, people are accustomed to carrying multiple devices that are connected to the Internet. In fact, in 1st world countries, you'd be hard pressed to find someone who doesn't own a cell phone. Being able to 3D print our own phone cases, tripods, camera upgrades and stands solves problems for a device we use everyday. The Square Helper created by Chris Milnes, which prevents the Square Reader device from spinning, is an excellent example of a mobile accessory business created from 3D printing. Chris found a need, albeit a very small need, and managed to create a profitable business out of this niche market.

Toys & Scale Models

The toy industry is undergoing an enormous transformation. Hasbro has even teamed up with shapeways to release a line of My Little Pony action figures. The mere fact that toy manufacturers are partnering up with 3D printing services means that even the toy industry knows it is undergoing a serious revolution. The industry stands to be disrupted by 3D printing because of the miniature and artisan nature of toys. With a 3D printer at home, board games, action figures, dolls, building bricks and scale models can all be easily designed, printed and shared.

While 3D printing opens up new mediums for aspiring toy designers, existing toy companies will have to creatively utilize 3D printing within their existing business models, possibly by offering customization as a service or partnering up with innovative designers.

Jewelry

Jewelry is a dazzling field of possibility for custom creations. The coveted nature of precious metals and intricate body adornments makes new designs a constant need for cultures all over the world. Right now jewelry creation with 3D printing involves printing in wax so the model can be casted. In the future, once 3D printing metal becomes a viable option, the industry (along with other industries that use metal heavily) will swell. Nevertheless, designing jewelry with CAD and 3D printing is already transforming the way the industry is designing and making jewelry. Rings, necklaces, bracelets, anklets, earrings and accessories are just a few jewelry categories ready for change.

Clothing

Reinvent the way humans wear clothes. Play around with 3D modeling to invent innovative wearable shapes. Use new materials to produce interesting fabrics; textures like chainmail can be created from interlocking patterns. 3D printing is excellent for producing outlandish original clothes for cosplay or fashion shows. Many costumes for films are created with 3D printing such as the suits for The Incredible Hulk, The Avengers and Iron Man. Shoes, shirts, headwear, eyewear and dresses are all ripe for innovation. I guarantee hundreds of clothing companies will spawn from the introduction of mass customization with 3D printing.

Religious & New Age Items

Oftentimes people find it difficult to instill belief and faith within themselves without hope of a greater being out there. Many times religions and churches are created for this purpose (and for money – the business behind religions like Catholicism or Scientology is massive). If you have ever considered creating a religion, 3D printing is definitely a viable way to prototype your religious artifacts. Believe it or not, the Internet and the information age has resulted in the creation and sharing of many 'New Age' items. Given their high price points, 3D printed metaforms are exceedingly easy to design and create with CAD programs. Since most large big box retailers, both offline and online, have banned 'magical' items, a tremendous opportunity exists for expanding religious beliefs new and old.

Tools

A 3D printer is a tool with the ability to make other tools, which makes it the ultimate tool. Measuring cups, screws, bolts, hinges, locks are all easily replaced with 3D printing. Everyday tools such as screwdrivers, wrenches, calipers and measuring instruments can all be printed on demand. An engineer in the field or in a remote location will be able to print his or her tools on demand, efficiently saving time and resources. Indubitably and ironically, we will witness the rise of new inventions that are first made with 3D printing.

Instruments

Personally, one of the coolest potential applications of 3D printing is making instruments. With material no longer a barrier for shape, imagine the kinds of instruments that can be fabricated from odd shapes. Possible ideas include percussion instruments such as the ocarina and the flute. Other instruments like the violin have also been 3D printed, so anything is possible. Add extra holes. Remove shapes. Make new instruments and create your own sound. It may be an ugly device, but does it make beautiful music? Musical creation is just another art form and 3D printing is a conduit to this craft.

Repairing, Replacing and Reinventing Parts

A 3D printer is the ultimate hobby replacement and creation solution tool. Those with expensive interests such RC, auto, drone, plane and yes, 3D printing connoisseurs, can now print and test their own parts rather than purchase replacements from a hobby store. Testing a new wing or chassis design can be done within the confines of a home. RCs and Drones can be repaired and upgraded slowly. Potentially, all of these items can be made with a 3D printer.

Scale models of all shapes and sizes are perfectly positioned to be 3D printed. Combining 3D printing with Arduino to create prototypes is a common practice today especially in the robotics and wearable technology fields. With these two technologies robots, generators and vehicles are all within a realm of possibility. The blueprints of a design can exist within .stl files and Arduino code and then be shared across the world with others. Don't let mechanical design limit your dreams when there is a world of programmable open source electronics (much easier to learn than Electronics of yore) to be explored as well.

Artifact Restoration & Rare Items

What if the tomb of Tamerlane were not stuck in Samarkand? Timur was a Turko-Mongol conqueror that reigned over the Caucasus, Mesopotamian, and Central Asian regions in the

1300s. His mastery of the Persian, Mongolian and Turkic languages gave him considerable understanding of cultures in the territory. Timur's armies were also largely multi-ethnic and he relied on Islam to legitimize and propel his power. Fascinated by Persian culture, Timur took it upon himself to introduce art, literature and architecture into the culture of his empire. Timur is regarded as the last great nomadic conqueror of the Eurasian steppe (before the advent of guns). Even though Timur and his armies wiped out approximately 5% of the world's population and commanded Eurasia, the conquests of Alexander the Great and Genghis Khan overshadow his legacy. Unlike these two great rulers, Timur is often reviled in history for his heedless destruction of terror and gruesome mass slaughtering's. The Timurid Empire fell apart shortly after his death.

In 1942, Joseph Stalin demanded the opening of Timurs' coffin despite the warning inscription on the coffin, "The greatest war will happen if the grace of Timur is open". What followed a couple days later was Hitlers attack on the Soviet Union resulting in one of the deadliest wars in human history. The possibility of 3D scanning means archaeological items can be digitized and shared, so original artifacts can be left alone. Imagine if the Russian excavators had merely 3D scanned the tomb for replication rather than opened it?

Duplicating hard to find or one-of-a kind items such as fossils and paintings for educational purposes has an enormous benefit. Instead of reading about Brontosaurus fossils or Van Gogh paintings and consuming information slowly by word, teachers can print out a replica and show it to the class for them to hold and experience themselves.

Pets

The pet market has enormous opportunities for 3D printing. Scale of items definitely has to be taken into consideration, as it will be easier to print gerbil accessories than dog accessories for example. Get creative with your items. How about an Eiffel Tower for the fish tank? Or how about a new toy for your dog? Various pets are popular in different areas of the world and depending on locale, have different needs. There exists a very large niche market of underserved accessories for less popular pets such as spiders, snakes, turtles and crickets.

Food

This is the category I am most excited about. Imagine 3D printing a pizza with 3D printed cheese, bacon and pepperoni. Okay, maybe the picture of a plastic-esque textured pizza doesn't sound too appetizing today, but the ability to 3D print food has enormous benefits to society as a whole for example, world hunger. There are probably advancements in 3D printed vegetables, fruits and meats but I'm sure the focus is on grain, dairy, pastry, candy and chocolate foods. 3D printers will make custom edible creations a normal thing. Food 3D printers include the ChocEdge, which is a 3D printer for chocolate creations, the ChefJet that is able to print sugary creations and the Foodini that can create savory and sweet cuisine including pasta, pizza, burgers and chocolate.

Reach Your Niche Market On The Internet

"The product went from idea to the printer in a single morning before work, and 3 weeks later I had a product for myself and to share around the world." –Jonathan Bobrow, Founder, Bitwise

After the creation of these items, one must decide on the business model. How are you going to monetize your niche? Who is the target audience? How are you going to reach them? These are some of the questions you will need to ask yourself.

The Internet is ***the*** way to reach a global market. The World Wide Web has already matured into a predictable economy and is now a catalyst for 3D printing growth. Building a website is a much simpler task than it was in 1995 or even 2005. Services like **Wix** and **Square Space** allow people to easily create and design functional websites. Those who want customization and refinement will find **Wordpress** to be the best option. All you need to get started with a website is a domain name and a host.

Find your customer online by simply locating the websites your target customers visit and reaching out to them. Searching related **Facebook groups**, **Google communities**, **Twitter hashtags**, **forum posts** and **blog comments** are great ways to directly reach your customer. From researching these networks, you will be able to find out your target customers needs, wants and hopes and create an avatar of your customer in the process. Be sure to post questions and answers to questions other people have. Helping others creates a reciprocal effect when it comes time to present your solution.

Gaining targeted traffic is a painless task. First off, write helpful and engaging content-rich blog posts that are easy to share. The result is a myriad of keywords that are indexed in Google, so anytime someone searches a keyword that is within your post they will have a chance of landing on your page. As your website ages, this is easiest way to rank higher on Google.

Niche exists because of the Internet. Before the Internet, you'd just be considered weird for having particular tastes. Now you can communicate and connect directly with others who share the same interests over the Internet on places like **Reddit**. You can buy anything you can think of online because there just isn't enough shelf space at Wal-Mart, the biggest superstore chain in the world to cover everyone's strange interests. Lastly, I propose creating a series of **YouTube videos** about your niche to link back to your website and get your physical item seen in cyberspace. It is the second largest search engine in the world and people love telling the cold-hearted truth on YouTube comments.

10 Places Online To Sell Your Physical Creations

NAME	URL	WEB RAFFIC (per month)
Amazon	http://www.amazon.com/	123,180,110
Ebay	http://www.ebay.com/	68,276,007
Craigslist	https://geo.craigslist.org/iso/us	57,571,068
Etsy	https://www.etsy.com/	18,356,804
Overstock	http://www.overstock.com/	14,038,718
QuiBids	http://www.quibids.com/	2,742,818
Buy.com	http://www.rakuten.com/	2,220,657
Oodle	http://www.oodle.com/	1,853,280
Bonanza	http://www.bonanza.com/	900,022
ClassifiedsGiant	http://www.classifiedsgiant.com/	763,945

DEFCAD

DEFCAD (or Defense Distributed) created by Cody Wilson deserves a special mention. The mission of DEFCAD is to defend the human and civil right to keep and bear arms as guaranteed by the United States Constitution and affirmed by the

United States Supreme Court. The Austin-based startup is often considered the "Pirate Bay of 3D Printing". DEFCAD offers a search engine and web portal for designers and hobbyists to find and develop 3D printable CAD models online. The website offers free information and knowledge related to the digital manufacture of arms. This is not an opportunity to highlight the ability to 3D print guns, but to underline the freedom of manufacturing and disruption that 3D printing enables for everyone.

The Power of Free for Big Businesses

A business can utilize the downtime of 3D printers to create giveaway 3D printed logos, actions figures or trinkets as gifts to go along with purchases. People love free things and oftentimes the thought will go a long way towards a repeat customer or social media share. Even if you are a solo entrepreneur, give things away for free. It is always an appreciable way to do business.

III. Build 3D Printers

"I find out what the world needs, then I proceed to invent. My main purpose is to make money so that I can afford to go on creating more inventions." –Thomas Edison

Building a 3D printer may seem daunting, but with due persistence and ingenuity it may just be the most rewarding method. Since the inception of Makerbot and its gradual decline there has not been as big of a dominant figure in consumer level 3D printers. Who will become the "Apple" of home 3D printers? A machine with the right combination of materials offered, size, quality, speed and price accompanied with a killer marketing strategy has the potential to succeed with this grand opportunity.

Remember, the PC market started promptly with a Microsoft monopoly and gradually but surely Apple took command with human interface brilliance and creativity. The market of building and selling 3D printers may be swarming, but there is a good reason for that. In an emerging market new competition always has a chance.

Democracy of Distribution, Manufacturing and Funding

Behind it all, the Internet really solved the problem of distribution and demand for both consumers and entrepreneurs. Finding out whether a market is viable is no longer a guessing mans game with communication and evaluation tools at our fingertips. Assessing global demand is as simple as searching trending topics on Twitter or asking a question on Facebook. Likewise, social networking platforms such as **Meetup** are tremendous for targeting similar-minded individuals and organizing physical hangouts. There are now hordes of methods to market a business, both online and offline. We are in a world of abundance and opportunity – just don't get overwhelmed by the possibilities.

The amazing thing is that the democracy of distribution and creation has literally come full-circle. Tools such as Logic Pro X, Kindle Publishing and even iPhones are redefining the way media is recorded and shared. Self-publishing services such as YouTube, Soundcloud, Apple App Store and Amazon allow creative types to publish media and share at will. The recent maturity of consumer level 3D printing and introduction of open-source programmable hardware is making it easier for people to become makers and subsequently, entrepreneurs. In the 21st century, everything man wants to create, he can easily create. With distribution and manufacturing no longer a roadblock, the last hurdle for hungry entrepreneurs is raising capital.

In 2012, Barack Obama signed the Jump Start Our Business Startups (JOBS) Act, which introduced equity crowdfunding as a cost-effective way for small businesses to raise capital, in turn complementing sites like Kickstarter and IndieGoGo who are leaders in the pledge for funding project space. The realization that people can easily raise funds for a project on the Internet has attributed to the rapid growth and popularity of crowdfunding. With crowdfunding stimulating an influx of makers and entrepreneurs, it is no surprise that some of the top funded projects on Kickstarter are 3D printers. To put 3D printer innovation in perspective, try and imagine where computers would be if Kickstarter existed during the 1970s.

The RepRap Project

The RepRap Project, created by Adrian Bowyer in 2005, is a free open-source 3D printer that has greatly attributed to the movement of low-cost and home-based 3D printers. The ultimate goal of the RepRep project is to be able to print itself. Each RepRap printer is simple and affordable to manufacture, making the RepRap the most widely used 3D printer by makers all over the globe. The blueprints of the RepRap project inspired early Makerbot 3D printers, the pioneer in desktop 3D printing. Prior to Makerbot and the RepRap project, 3D printers were industrial use only coming from manufacturers such as Stratasys and 3D Systems and costing upwards of $10,000.

Thanks to open source innovation the maker community has collectively been able to impact the rate of innovation of consumer friendly 3D printers subsequently decreasing prices and strengthening printer designs. If you search to buy a 3D printer for your home today you will no doubt find a plethora of consumers 3D printers available – largely owing thanks to the RepRap project.

The RepRap Forums are filled with FAQs, questions and answers from aspiring makers around the globe. Share your stories, insights and knowledge with the community, so they may share their wisdom back. If you plan to design and create your own 3D printer, the RepRap Forums is quite possibly the best 3D printer making resource online. Instead of rewriting what is already available online and doing less of a job, I encourage you to directly engage and learn from the RepRap manual:

https://reprapbook.appspot.com/

The first and most important step to getting started with a 3D printer business is to create the prototype. Marketing, distribution and manufacturing come second only when the first step has been completed, tested and refined. Creating compelling competitive advantages will be the main challenge that sets your 3D printer apart from the rest of the competition.

I recommend finding a team that will help guide the project in all aspects including Project Managers, Designers, Electrical, Mechanical and Software Engineers. Collaboratively, the team will need to decide on layout design, mechanical features and a strategy of focus. Those on your team must possess familiarity with 3D printing and have a curiosity to learn and improve.

Here are some things to keep in mind when designing the blueprints of a 3D printer:

- ❖ Size
- ❖ Design
- ❖ Material of printer
- ❖ Output material
- ❖ Quality of output
- ❖ Cost
- ❖ Cloud Compatibility

The features you decide to implement into your 3D printer will have a large impact in determining your customer. The size of the printer, the materials used for printing and extra features will narrow down the subset of customers you will be able to serve. For example, a 3D printer with a small build volume and high detail resolution is ideal for the dental or jewelry industries.

With Great Challenge, Comes Great Satisfaction

"The battle is not won by those who are swift, but by those who endure"

Even if you don't plan on selling your 3D printer, building one is a rewarding challenge. For starters, consider purchasing a DIY 3D printer – in the long run it will save you both in costs and technical headaches. Building your own 3D printer is a worthwhile investment if only just to learn how a 3D printer works. I still remember when I bought my first 3D printer: a PrintrBot, possibly the simplest RepRap based DIY 3D printer there is. It took me a whole week to put the printer together, but when it was finished I felt a complete warm sense of accomplishment even though it was only capable of making plastic doo-dads. The quality of the 3D printer may have been primitive, but I built it and it worked!

Kickstarting Your 3D Printer

Kickstarter gives underprivileged artists and entrepreneurs a chance to share their ideas with the world, gain valuable customer feedback and potentially sizably fund their project. Kickstarter projects allow project creators to test the market without having to officially launch unless customer demand is validated. Creating a project also forces creators to produce media assets, which are cornerstone to building a brand/website anyways.

I encourage those who are interested in building and selling their own 3D printers to do their homework before selling on Kickstarter. Research and learn everything there is about your customers and the process of manufacturing 3D printers. Remember: Kickstarter is not the only option for showcasing and raising funds for your 3D printer. It is, however, a good launch pad for building media assets and accumulating customer feedback. Successful projects will garner a huge amount of interest and funding, but will encounter problems if forced to scale dramatically when it comes time to deliver.

For the sake of 3D printer enthusiasts everywhere do your due diligence before launching a Kickstarter project. Many successfully funded 3D printers have gone months past their original intended fulfillment date, resulting in frustrating anticipation from loyal and eager backers (including yours truly).

The following is a list of recommended assets to have completed *before* launching your 3D printer on Kickstarter:

- ❖ Working Prototype
- ❖ Bill of Materials
- ❖ Functional understanding of supply chain for parts
- ❖ Compelling project title
- ❖ Logo
- ❖ Video

Once these items have been successfully procured you'll be ready to launch on kickstarter. Proper planning ensures 3D printer creation success!

3D Printers Successfully Funded Through Kickstarter

3D Printer Project Name	Amount Raised
3DMonstr 3D Printer	$89,806
3DPandoras: The next generation in 3D Full-Color Printing	$156,885
AMAKER: World's First Dual ARM Open Source 3D Printer	$40,514 AUD
B9Creator - A DIY High Resolution 3D Printer	$290,150
B9Creator - A High Resolution 3D Printer	$513,422
BI V2.0 - A self-replicating, high precision 3D Printer	$210,830 CAD
Bukito Portable 3D Printer	$136,984
Bukobot 3D Printer	$167,410
Cobblebot 3D Printer	$373,916
CreatorBot-3D	$75,517
DeltaMaker: An Elegant 3D Printer	$152,597
Deltaprintr	$236,451
DeltaTrix 3D Printer	£19,350
Doodle3D	$73,777

FORM 1: An affordable, professional 3D printer	$2,945,885
Griffin Pro 3D Printers	$51,620
Helix 3D Printer	$125,140
HYREL 3D Printer	$152,942
Kossel Clear: A full-sized delta 3D printer	$266,337
LittleRP - Affordable Flexible Open 3D Resin Printer	$118,923
MakerLibre Delta 3D Printer v2	$43,969
Mamba3D	€ 39,785
MM1 Modular 3D Printer	$59,280
OpenBeam Kossel Pro - A new type of 3D Printer	$122,016
Pegasus Touch Laser SLA 3D Printer	$819,535
Phoenix 3D Printer	$109,563
Printrbot: Your First 3D Printer	$830,827
Printxel 3D Printer Beta Kit	$12,077
QU-BD One Up 3D Printer	$413,530
RepRap Open Source 3D Printer	$19,535 CAD
RepRap: The Self Replicating DIY 3D Printer	$3,961

in Your Home	
RigidBot 3D Printer	$1,092,098
RoBo 3D Printer	$649,663
Robox : Desktop 3D Printer and Micro-Manufacturing Platform	£280,891
RoVa3D : The First 5 Material/Color Liquid Cooled 3D Printer	$132,120 CAD
Solidator DLP Desktop 3D Printer	$144,403
The Buccaneer®	$1,438,765
The gMax 3D Printer	$129,224
The Micro: The First Truly Consumer 3D Printer	$3,401,361
The Peachy Printer	$651,091 CAD
The Vision	$65,346
Ultra-Bot 3D Printer	$45,540
ZEUS: The World's First ALL-IN-ONE 3D Printer / Copy Machine	$111,111
Zim, the true Consumer-oriented 3D printer	$347,445
Zortrax M200 - professional desktop 3D printer	$179,471

IV. Share Your 3D Printer

"The Factory In The Cloud"

Realize that a quality 3D printer is not cheap and won't become a staple of society like computers are for years, possibly decades to come. Consider your 3D printer as an asset just like you would for a home or car.

Monetize the capacity of your 3D printer by printing things for yourself and for others. A 3D printer has a defined rate of output, so the maximum printing capacity of your 3D printer has a set amount for each day. In a lean system, an idle system is a form of waste. Why not maximize the printing ability of your 3D printer? The output of a 3D printer can be calculated with the following formula:

$$Rate\ of\ output = \frac{Amount\ of\ Material\ Printed}{Time\ Spent}$$

By increasing the amount of material printed, the rate of output is effectively optimized.

The Shared Economy

The concept of a shared economy is a relatively recent notion. Ride sharing services such as Uber and Lyft are disrupting both the Taxi and freelancing industries. Airbnb gives those with a spare room or an empty home an opportunity to make extra money. The development of a shared economy is effectively optimizing the efficiency of the resources on our planet Earth. We might as well share our 3D printers too.

3D Printing Hubs

Visualize yourself as an aspiring watch designer headed to Paris from Los Angeles for a pivotal business trip, and right before takeoff your co-worker calls and reminds you to make a slight design modification. With hardly any time to spare, you are forced to re-3D design your product on the flight. Upon landing you immediately locate and send your design to the nearest 3D printer in town. The upgraded watch design is printed right before the meeting and you are rewarded with a large contract. This is a great example of just-in-time manufacturing and the wondrous possibilities of an integrated global 3D printing network.

What if you could help 3D print other people's ideas for them bringing *their* dreams to life *and* get paid? Well, you can. Essentially, you can literally print money! Networks of 3D printer systems give makers a plethora of options when it comes time to choose material and detail. Access to a selection of 3D printers is difficult to come by so the time and material usage of your 3D printer is actually very valuable. Consider investing in a quality 3D printer and putting it up onto a 3D printing network to maximize its printing capacity.

The 3D Printing Networks

There are really only two major 3D printing networks – 3D Hubs and MakeXYZ.

3DHubs

On 3D Hubs there are 10,000 different 3D printers available in more than 80 countries. With a massive and growing user base if you are in a metropolitan city then 3D Hubs is an excellent place to share your 3D printer. You have direct communication with the 3D designer and will choose between meeting face to face and sending your prints through mail. The largest base populations of 3D printers around the world are in New York, Milan, London, Amsterdam and Los Angeles. This density of selection for certain parts of the world is what makes 3D Hubs an ideal place to share your 3D printer. There are makers all over the world and both established and developing countries are innovating with 3D printing. Remember that you still incur the costs of 3D printing materials, maintenance, machine wear and tear and electricity. I encourage you to look at the Trend Report on 3D Hubs – the data will definitely enlighten you on the growth of 3D printers around the world.

MakeXYZ

MakeXYZ allows makers to find a local idle 3D printer and print their brilliant designs. It would be a good idea to list your 3D printer on both 3DHubs and MakeXYZ to maximize exposure. MakeXYZ calculates pricing based on the volume of the part file. The materials used for the support structure is not calculated by MakeXYZ. The pricing also does not include machine time or set up fees. Both MakeXYZ and 3DHubs take a whopping 15% commission, but they add it to the price you charge (you don't lose money, the customer pays more).

Fiverr

With Fiverr creativity is the only limit to what tasks people can perform for others. For $5 a task, people are willing to perform all sorts of assignments for you. Use Fiverr to share your 3D printer, print stuff for other people and their businesses and possibly even share your 3D modeling skills. Potentially, Fiverr can be used to monetize your 3D printer in all 7 ways listed in this eBook.

The differences between each 3D printing network are size and type of community for makers and sharers, selection of 3D printers, pricing models and the usability of the site. When looking at a 3D printing network these factors will all need to be taken into account. The 3D printer you choose is also pivotal factor for the quality of your prints. In the last few years, desktop 3D printers have improved significantly, but not all of them can reproduce excellent results. Since there are so many new 3D printers released every year, I present to thee, a guide for 3D printers in the new year!

2015 3D Printer Buyers Guide

Ultimaker 2

The Ultimaker 2 is still one of the easiest and most reliable 3D printers available on the market today even though it was released last year. It ships completely preassembled and comes with Cura, Ultimaker's own open-source software. With a large build volume, heated glass bed, onboard OLED controls and illuminated interior the Ultimaker 2 is a feature rich 3D printer for all skill levels. Buyers looking for a high quality, ease to use 3D printer will do themselves a favor by picking up an Ultimaker 2.

Price: $2,499
Build Volume: 230 x 225 x 205mm

Lulzbot Taz 4

The Lulzbot Taz 4 is the latest evolution in the line of versatile `Taz printers that are based off of the RepRap engine. The quality and consistency of this plug-n-play open source device has improved dramatically from previous versions. The Taz 4 has a large heated glass bed, which allows for big prints in a wide variety of filament materials. The design of the 3D printer won't be winning any awards soon, but it is a production tool, created by engineers for engineers. This is the perfect 3D printer for makers, hackers and engineers.

Price: $2,195
Build Volume: 298 x 275 x 250mm

MakerBot Replicator (5th Generation)

The newest version in a pioneering line of 3D printers, the MakerBot Replicator is gem of a 3D printer with boatloads of features. Sporting an LCD interface with a sexy precision knob, cloud compatibility and an on-board camera makes the newest Replicator a powerful creation tool. The

MakerBot Desktop software is incredibly easy to use and the prints are high quality as well. The MakerBot Replicator 5th generation is a quality 3D printing tool for all skillsets. While engineers and makers may have richer options, those looking for a high quality 3D printer to show off to friends that is easy to use can really not go wrong with the Replicator.

Price: $2,899
Build Volume: 252 x 199 x 150mm

Ditto Pro

The Ditto Pro with its silky white finish and clean cuts is an aesthetically pleasing performance 3D printer. If anything, the unique C-shaped design sets it apart from the rest of the 3D printers. The machine has both an interchangeable extruder and print bed, a bright on-board graphic display and an SD card input slot. Not only is the print quality highly impressive, but the Ditto Pro is affordable. The Ditto Pro is designed for professionals and consumers, but is ideal for tinkerers.

Price: $1,899
Build Volume: 220 x 165 x 220mm

Printrbot Simple Metal

The Printrbot Simple Metal is a fun assembled 3D printer that has been upgraded significantly from previous versions. Its black powder coated all-metal construction will look slick on your desk next to your laptop and lamps. The bed is now warp resistant and leveling is simplified now with an auto-leveling probe included. Although the first Printrbots were entry-level DIY 3D printers, the Printrbot Simple Metal is a solid professional choice for an affordable price. Highly recommended for the budget conscious 3D printer shopper.

Price: $599
Build Volume: 150 x 150 x 150mm

BeeTheFirst

The BeeTheFirst 3D printer is a user-friendly printer with smart design. In fact, the machine has cleverly fused ergonomics and design in its portability. The leveling bed and spools are magnetically held together. The software is perfect for beginners and advanced users will find a repository of open-source software. The portability and usability of the BeeTheFirst 3D printer makes it a great choice for all users and skill levels.

Price: $2,171
Build Volume: 190 x 135 x125mm

Zortrax M200

Don't let the alien name scare you, but the Zortrax M200 will print some divine things for you. Its large build volume provides for a superb printing experience. The aluminum body of the Zortrax M200 is engineered for durable prints. Although the software doesn't allow for printing temperature control, the perforated bed and leveling system help to create a complete quality print. Overall, the Zortrax M200 is a fine machine that can print large and accurate prints.

Price: $1,990

Build Volume: 200 x 200 x 185mm

Idea Builder

The Idea Builder from Dremel is the first 3D printer created by Dremel and is a product targeted towards the everyday consumer. It looks like a home appliance that you would have in your garage (or kitchen). This 3D printer is easy to get up and running. With an in-depth quick start guide, the Idea Builder is as simple to use as most of Dremel's products have been. Price point, quality and host of features make the Idea Builder an ideal consumer friendly 3D printer for makers of all ages.

Price: $999
Build Volume: 230 x 150 x 140mm

Form 1+

Form 1+ is the upgrade over the original Form 1 from Form Labs. This time the sophisticated stereolithography printer feels and is high quality. The overall design is s suave and the prints that come out are some of the best you will be able to find in a consumer 3D printer. The prints come out faster and in even better quality before. Form Labs printers use stereolithography, which is a form of 3D printing that uses a laser to cure resin into a 3D shape as opposed to fused filament deposition where PLA or ABS plastic is layered on top of each other. The upgrades in the Form 1+ printer makes an unparalleled printer for those who want to print in high detail resign. The Form1+ 3D printer is ideal for those who are unyielding and only invest in high quality and precision.

Price: $3,299
Build Volume: 125 x 125 x 165mm

ProJet 1200

The ProJet 1200 is a professional machine targeted towards those who work in a small scale and use investment casting. The ProJet 1200 is also a stereolithography 3D printer, but how it differs from the Form 1+ is that the resin output can also be castable resin. Simple to use onboard controls and dependable quality makes ProJet

1200 an excellent professional printer. Its smaller highly detailed build volume makes it limited for those who work under such conditions like jewelers or dentists.

Price: $4,900

Build Volume: 43 x 27 x 150mm

Factors To Look For When Buying A 3D Printer That Will Affect Your Rate of Output:

- ❖ Quality of output
- ❖ Printing material
- ❖ Size of 3D printer
- ❖ Cost
- ❖ Speed

If you plan to share your 3D printer as a means to make money, choose a good 3D printer and maintain the above factors to optimize your output.

Expanding The Distributed Factory

Looking ahead, after acquiring multiple printers one could open a 3D printing store or Maker Space. If your niche business is doing well, acquiring more 3D printers may be the best way to scale up. These printers can make even more money by joining a 3D printing network. The potential to offer classes and start a small mass manufacturing business on demand is there. This route can easily be combined with chapter 7.

V. Customization as a Service

The birth of the customization on demand era begins with **3D printing**. **Programming** and **Automation** allow creators to leverage 3D printing technology to offer made-to-order products. The ability to modify designs and manufacture with a computer translates to manufacturing that can be automated and customized on demand. Customers will be able to personalize what they want on a website or application while computers and 3D printers perform the labor of creation.

The Potential Of On-Demand Mass Customization

- ❖ Production is based on demand
- ❖ No inventory
- ❖ Low overhead cost
- ❖ Automation
- ❖ Design flexibility
- ❖ Integrated Cloud Networks

By eliminating human labor for manufacturing processes (which is oftentimes dangerous and dehumanizing in countries with loose labor laws), these workers can be utilized for more efficient work such as design or marketing.

Case Studies

Shapify

Shapify is a custom service that offers 3D printed mug shots. Have you ever wanted to create an action figure or sculpture of yourself? Shapify can make your dreams come true by 3D scanning your body and creating a mini 3D printed version of yourself. This technology doesn't just stop with human caricatures; it has potential for living creatures and inanimate objects of all varieties. Narcissists and narcissists who want share their faces to friends, keep an eye out for the nearest Shapify scanner.

Feetz

The customization process of Feetz is simple. Customers download an app on their smartphone or tablet and then they take photos of their feet. The software custom designs a shoe that appeals to them. Buyers provide data such as height, weight and other lifestyle preferences and a perfect pair of form fitting shoes is shipped to them. The shoes arrive at the customer's door within 7 days. Feetz is truly customization on demand.

Companies Creating Customization As A Service

Company	Business
MakieLab	3D Printed Dolls
Tecnologia Humana 3D	3D Printed Fetus
Spuni	3D Printed Spoons
Make Eyewear	3D Printed Custom Eyewear
Protos	3D Printed Custom Eyewear
Twindon	3D Printed Figurines

Twinkind	3D Printed Figurines (highly detailed)
My3Dtwin	3D Printed Figurine service
Sexy Objects	3D Printed Sex Toys
Mixee Labs	3D Printing Customizable Products

Computer-Integrated Systems of Planning and Design In Manufacturing

Creating a fully integrated mass customization on demand service requires a fruitful combination of programming, automation and 3D printing. Manufacturing customized products is possible with a computer-integrated system of 3D printers and software. Building a robust system for physical customization takes a team of Software, Mechanical & Electrical Engineers, Operations Experts and Designers.

Success requires choosing a niche and validating that there is appropriate demand before creating anything. Will people want to customize this item? A couple Google and Amazon searches will tell you plenty about what people are buying. Building software that integrates customizability with 3D printing as an online service or app will take a team of specialists and engineers. The layout of customization design will need to be designed for usability. If you have a clearly outlined business model and enough funding then creating a customization service is a highly viable option.

Partnering with 3D Printing Service Providers

There is the option to partner up with a 3D printing service and use their equipment & facilities. Shapeways, iMaterialise and Sculpteo each provide their own API's for 3D CAD tools and creation apps. The 3D printing service provider does the printing, polishing, packaging and customer service for you. You receive a royalty for every design someone else prints using your app. **Solid Concepts**, among other similar companies, provides a custom manufacturing service for scales of all sizes.

Ponoko for Prototyping

Ponoko is the world's first online service for creative people to use online manufacturing service for prototyping. With 3D printers, laser cutters and CNC milling machines, Ponoko differentiates itself from pure 3D printing services like shapeways. As one of the first services to offer distributed manufacturing and on-demand manufacturing, Ponoko is at the forefront of a paradigm shift in manufacturing. If your prototype requires a synthesis of materials such as wood, metal and plastic and a blend of cutting and creating tools then seriously consider Ponoko for your next project.

"Creative crafts will be of great value for the economy of the future. It will answer the growing demand for quality, creativity, and authenticity."
–Klamer

4 Steps To Bootstrapping A Niche Business On The Internet: Creating the Minimum Viable Product (MVP)

Starting a business on a budget by bootstrapping your resources towards a niche business online is much simpler than one would think. A website is the easiest and most common way for a customer to learn about you so far as to even say it is the "face" of your brand. By keeping a simple layout you can save thousands as you grow your business. Having a blog is very important for communicating to your customer that you are not some corporate giant, but that you understand your target audience. A blog helps you relate to your customer on a personal level and for them to reach you.

Step 1: Create blog.

Creating a blog is **extremely** streamlined. No HTML or coding is required. In fact there are so many options available that it will probably confuse you before helping you. It is affordable, easy to set up and a great way to gain a lot of followers quickly. Getting started only takes 5 minutes. Pat Flynn, from Smart Passive Income, shows you how you can build a blog within 4 minutes. Make sure to maintain your blog and periodically update it with new blog content. Search engines love fresh content, so the more often you post the easier it will be to gain traffic.

Build a Blog in 4 minutes:
http://youtu.be/wPwQvnar99w

Create a blog (Time: 5 minutes – Cost: Under $10)

Step 2: Traffic

Write blog posts with compelling headlines. Relate the benefits of your product into helpful how-to or list blog articles. Depending on your niche, I recommend choosing one of either Facebook or Instagram to build your visibility and marketing. It is easy to run contests and promotions. Both services are free and have massive user populations to tap into. Contact other blogs and volunteer to guest post and this will drive targeted traffic from their blog onto yours.

Quality Blog Posts and Social Media

Step 3: Sell

Put your 3D printed niche items up for sale. The beauty of having your own web store means that you control the prices, placement and marketing of your own items. Start with one item as you get everything set up. Consider creating information products such as Tutorials, Videos, eBooks and Podcasts.

Take pictures of your item, add a description and put it up for sale

Step 4: Monetize

A website doesn't only have to make money from sales of products. If your website generates a lot of traffic because of popular content and blog posts then consider partnering up with affiliate or ad networks. Contact businesses directly and ask if they would like to purchase an ad space on your website. Every website can be fine tuned in a way so that ads look natural and link to helpful products related to your niche.

Monetize – Google AdWords, Affiliates, Sell Ad Space

An important thing to keep in mind: Do not rely on social media to create your brand. You do not own assets on these websites and they do change their Terms Of Service at will. You can sell a website, but you can't sell an Instagram profile, unless it is included with the sale of the main business which just further emphasizes this statement. The methods I listed in Chapter 2 for finding your audience, "Reach Niche Market Through Internet Research & Networking" are great for communicating with and garnering traffic. If you are the target audience then it is much simpler to reach those in your demographic.

Subscription Business Model

Commit to obtaining subscribers and create new 3D printer models and/or products on a regular interval to give away. Ask for subscribers instead of donations. Consider adding information products to complement your business model. With a list full of people interested in what you do there are many ways you can market yourself, especially for when you do go for the sale.

Here Is A Neat Ruse To Building Passive Income:

With your niche in mind, create a table of contents for a book you would like to write. This can be something as simple as a how to guide or a collection of information.

Write a blog article for each topic in the table of contents. Take your time and create quality content. Most websites take 6+ months to properly index and rank well on Google for blog posts, but continuous posting will translate to superb **Search Engine Optimization** (SEO).

Compile the collection of quality blog posts and turn it into an eBook. A collection of topics compiled into one is far more valuable than a blog post on a random topic. Publish on Kindle.

Rinse and repeat.

If you are not a writer or despise writing, consider investing in a high-quality microphone and speech to text recognition software. On the off chance that you love seeing yourself on camera, you may as well acquire high quality video equipment too. With this method you can create YouTube videos and podcasts that are translated to blog posts that are then compiled into eBooks.

P.S. You can create an app and monetize information as well.

VI. Invest in 3D Companies

"The next trillion dollar industry" – Business Insider

As the 3D printing industry grows, the major 3D printing stocks will certainly grow too. Remember, investing in stocks is a passive process, which, depending on how you view investing, can be a good or bad thing. Although the 3D printing industry is *expected* to achieve $10 billion in sales by 2020, stocks are often viewed as a volatile investment especially after the recession, so use your best judgment before beginning stock trading.

The great mystery of the 3D printing technology is how its vast potential will develop and revolutionize the economy. There are already many ways to make money with 3D printing, but what new possibilities will open up in the future? The potential of the 3D printing industry is substantial so it is important to highlight key stocks. Like any other serious venture, it would behoove an investor to learn the fundamentals of the stock market before trading.

Putting Money Where My Mouth Is

I have personally owned both Stratasys and 3D Systems stocks. I purchased both near the beginning of 2013 and sold them in the summer of 2014. I sold both of these key stocks to fund other investments (such is the life of an entrepreneur where time is a crucial currency). I did make a decent return ($) over the 15 or so months.

I do not proclaim to be a fortuneteller. As a 3D printing enthusiast and entrepreneur, I do my best to be well prepared for trends in the stock market. Just know I couldn't honestly recommend anything without first believing in it myself.

3D Printing Stocks That Are Disrupting The Industry

Stratasys (SSYS)

In my opinion Stratasys is the best-positioned 3D printing stock due to its dominance in both industrial and consumer 3D printing. In 2013, Stratasys acquired Makerbot making it a brand leader in consumer 3D printers. Stratasys' multifaceted line of Objet printers, more than $500 million in revenue in 2013 and widespread analyst approval makes them a leader for the future of 3D printing technology.

3D Systems (DDD)

Maybe the more popular name right now, 3D Systems Corporation also boasts a strong line of industrial and consumer 3D printers. The company offers 3D printing solutions, services, parts, scanners and education resources. The stock did take a hit this year; so depending on entry point DDD could be a strategically positioned stock. If you are looking to capitalize on the 3D printing industry as a whole, take a long hard look at 3D Systems.

Always be wary of stocks with a small market capitalization and high risk. Perform in-depth research on new technologies because these will be the biggest catalysts for growth. The following 3D printing stocks are all growing players in the 3D printing industry ranging from 3D printed metals to 3D printed tissues:

VoxelJet AG (VJET)
ExOne Co (XONE)
Materialise NV (MTLS)
Arcam (AMAVF)
Organovo (ONVO)

Stocks benefitting from the 3D printing revolution are not limited to just 3D printer makers and service providers. Existing companies that provide software solutions or 3D printing parts are also well positioned to gain from this industry. Many of these businesses will actually utilize 3D printing to gain a competitive advantage. The industrial uses of the technology, not just the consumer benefits, need to be kept in mind to predict how the technology will generate a boon in their stock value.

Quality Stocks Related To 3D Printing:

Hewlett-Packard Company (HPQ)
General Electric Company (GE)
AutoDesk, Inc (ADSK)
Adobe Systems Incorporated (ADBE)
3M Co (MMM)
Microsoft Corporation (MSFT)

In the relatively short term (1-3 years) all of these stocks are somewhat overvalued, but when viewing them long term (3+ years) the potential increase is almost limitless, especially as new scientific, medical and industrial discoveries are uncovered one by one. Positive advancements usually lead to stock price *jumps*. Since 3D printing stocks are still quite volatile, the best strategy would be to invest when there is a large dip in stock price.

Looking at the technology, it is obvious that it will have a large impact on our world, to what extent we are unsure but therein lies of the beauty of 3D printing as an emerging technology. **I am very bullish on 3D printing**.

VII. Teach

"The acquisition of wealth is no longer the driving force in our lives. We work to better ourselves and the rest of humanity." –Capt. Jean-Luc Picard

Sure, there are people who enjoy learning about 3D printing (like those reading this eBook), but the other 99% of the population will hardly care. Chances are high that others just want to use your 3D printer. Or bug you to no end until you 3D print something for them. For every person that can use a 3D printer there will be three more asking him or her how to use it. If you enjoy helping others and love 3D printing, why not teach others about this fascinating technology?

In this age of democratized everything including education, one can exploit his or her knowledge and teach others, particularly 3D printing. For example, Udemy is a site where independent instructors create and sell courses to students. This is useful for those who want to learn a specific skill rather than a broad general subject at a college. The growing elderly population will be clueless as to how to use a 3D printer, just like the Internet and computers before it and will seek help from others. Not everyone is willing to spend money to find out if they will use a machine or not.

3DPrintBoard is the ultimate discussion forum on all things 3D printing. In the last couple of years I have seen the community grow extraordinarily from a few hundred members to thousands. Remember how early we still are in the overall evolution of 3D printing. **Bld3r** is a 3D printing social network that lets you share 3D printed objects, tutorials, articles and torrents. Bld3r's Pinterest/Reddit styled layout is very practical for sorting through 3D printing related shares and commenting with the community. Feel free to ask questions too! These are great resources for learning what the 3D printing community does and loves.

There will be a demand for education with all 3D printing topics including:

- ❖ 3D Modeling

- ❖ Preparing Models for 3D Printing
- ❖ How To Use 3D Printers
- ❖ How To Build 3D Printers
- ❖ Opening a Maker Space
- ❖ How To Make Money With 3D Printing

Start small. Consider starting a YouTube channel similar to Khan Academy to teach others on their computers. YouTube is the world's second biggest search engine; so visual learners will certainly find you.

Whatever it is you want to do, the most important thing is to *do it*, and when you are doing it, *don't give up*!

There is a great joy to be found in teaching others. Not only does one learn better when educating others, but also teaching is a service that empowers others with constructive knowledge & ideas. People will pay to learn marketable 21st century skills. Lead the learning revolution!

Conclusion & Thoughts

3D printing is a disruptive technology of mammoth proportions, with effects on energy use, waste, customization, product availability, art, medicine, construction, the sciences, and of course manufacturing. We are entering an age of abundance. Anything man wants to create, man can create. The problem will be sorting through all of the possibilities and options.

As more and more tenacious creative people start businesses around 3D printing, the faster innovation will grow. Be early to this shift, because it *is* coming.

About The Author

Jeffrey Ito is a graduate of University of Southern California with a Bachelors of Science in Industrial & Systems Engineering and an SAP specialization. He has been a hardcore 3D printing enthusiast since the first time he laid eyes on a 3D printer in action. In his spare time, Jeff enjoys traveling, playing basketball, lifting weights and hanging out with friends.

To hear about Jeff's new books first, sign up to his New Release Mailing List

Thank You

I hope you enjoyed this book as much as I did writing it! If you have any questions, comments or think I have left something important out, feel free to personally send me your email: ito.jeffrey@yahoo.com

If you have an extra second, I would love to hear what you think about it. Reviews are gold to authors! If you've enjoyed this book, would you consider rating it and reviewing it on www.Amazon.com

www.ingramcontent.com/pod-product-compliance
Lightning Source LLC
Chambersburg PA
CBHW051712170526
45167CB00002B/634